FIVE WORDS
— TO SAVE —
THE WORLD

Peter Cockrum

Copyright © 2021 by Peter Cockrum. 831091

All rights reserved. No part of this book may be reproduced or transmitted in any form or by any means, electronic or mechanical, including photocopying, recording, or by any information storage and retrieval system, without permission in writing from the copyright owner.

To order additional copies of this book, contact:
Xlibris
AU TFN: 1 800 844 927 (Toll Free inside Australia)
AU Local: 0283 108 187 (+61 2 8310 8187 from outside Australia)
www.xlibris.com.au
Orders@Xlibris.com.au

ISBN: Softcover 978-1-9845-0767-9
Hardcover 978-1-9845-0768-6
EBook 978-1-9845-0766-2

Print information available on the last page

Rev. date: 09/01/2021

*For
Elise and Lyndal
Rory, Grace, Oriel and Andre
Bernard, Christopher and Kathryn*

Claude

My Mentor - Claude Culvenor PhD, D Phil, D Sc

PROLOGUE

The five words come together as two phrases:

"I believe"

And

"In my opinion"

What follows is an archipelago of ideas pushed up out of the sea of knowledge by musings born of seventy-five years of living.

I anticipate that many, like mariners of old, will pass by at a distance cautious of hidden perils, but I hope that some may, like Charles Darwin or Galileo Galilea, pause and ponder.

ON WORDS

TRUTH (Noun: A fact or belief that is accepted as true) (Soanes, Catherine (Ed.), 2003)

FACT (Noun: A thing that is indisputably the case) (ibid)

BELIEF (noun: A feeling that something exists or is true, especially one without proof) (ibid)

IDEA (noun: A mental impression) (ibid)

In my opinion the key point here is acceptance, and the responsibility of all of us to consider what we are accepting. When something is presented to us as being true it is incumbent on all of us to reflect on whether it is a fact or a belief. Furthermore, I believe great conflict can be avoided by the presenter nailing their colours to the mast up front and identifying their truth as a fact or a belief.

FACTS

The above definition of a fact includes the term "indisputably". Straight away we are on thin ice.

Since the turn of the 20th Century the organised body of knowledge known as science has undergone a massive upheaval. In the fields of physics and cosmology at least, use of the term 'indisputably' has become perilous indeed, but in mathematics it still has a secure place.

Mathematics, it seems to me, revolves around the process of proving that sets of information from different sources and forms of description can be shown by logical manipulation to be equivalent. Thus, a peer reviewed mathematical proof of something is as close as one can get to being indisputable.

Whilst mathematics underpins advances in fields such as Relativity and Quantum Mechanics, the ideas they spawn are philosophical, and the conclusions they seem to suggest are far from indisputable. Rather they represent the frontier of the inquiring human mind in pursuit of the understanding of ourselves and the cosmos we inhabit. Thus, we need to differentiate between the measurements we make which, conscientiously performed, can yield facts, and the conclusions we draw from them.

IDEAS

The definition above calls them mental impressions which is well worth some exploration.

We humans have five senses through which we interact with the world around us. Five types of sensory cells initially respond to chemical or mechanical stimulus then turn those responses into electrical signals which a variety of neurones transport to specialised processing regions in our brains which then create within our consciousness, what we refer to as smells, sights, sounds, tastes and textured surfaces. Importantly these are not the actual thing we are investigating, they are impressions, pictures in our brains, of what is actually out there; physically and temporally removed from the reality of existence

The temporal delay is on the order of microseconds (10^{-6} s) but I have written before (Cockrum, 2020) that there is a vast difference between this range and the Planck time scale of femtoseconds (10^{-15} s) at which quantum mechanics operates. Our impressions of reality around us are, in fundamental temporal terms, far removed from the events we think we are observing in 'real time'.

The physio-chemical responses to 'reality' are even more complex. For me, the most easily absorbed explanation of this complexity comes from the American author Neal Stephenson (Stephenson, 2008, pp. 374-380) in his book, Anathem. He describes what he calls in this book a 'calca' titled "The Fly, the Bat and the Worm". He hypothecates a quintessential Fly (all eyes), Bat (all ears) and Worm (all feeling) and speculates how they might communicate to, say, warn each other of an existential threat. Eventually he concludes that the only common language they could possibly use to meaningfully communicate their observations to each other is three-dimensional geometry: their only common language. He goes on to relate the Fly, Bat and Worm to our senses of sight, hearing and touch and speculate that we humans are an extension of these three creatures and such communication between such disparate inputs is a routine activity occurring in our brains: Consciousness.

Consciousness is where we create mental impressions: Ideas

BELIEF

Defined (above) as a feeling (without proof) that something is true or exists.

Given this definition, I think that belief is enormously powerful. It underlies our drive to explore all we perceive around us, in that we believe there is a structure and logic to our cosmos, minute and astronomical, which we are struggling to understand.

At a fundamental level, marrying Einsteinian Relativity and Quantum Mechanics is a coalface in our understanding of the structure of the universe, but we do believe that such a reconciliation is possible, and we speculate as to how to achieve it.

In the context of the Fly, Bat and Worm 'calca' referred to earlier, one candidate reconciliation, referred to as ' String Theory" works reasonably well but only if the universe is structured in 10 (or11) dimensions. Ouch! Awkward! Strains belief?

Well, mathematicians are not constrained in their thinking like physicists are. In his book: The Shape of Inner Space (Yau, 2010), Shing-Tung Yao, mathematician, topologist and winner of the Fields Medal[1], describes how serendipitously, as a theoretical exercise analysing the work of Eugenio Calabi, they had already developed a theoretical mathematical object called a Calabi-Yau Manifold – an 8-dimensional surface!

---

1   The **Fields Medal** is a prize awarded to two, three, or four <u>mathematicians</u> under 40 years of age at the <u>International Congress</u> of the <u>International Mathematical Union</u> (IMU), a meeting that takes place every four years.
The Fields Medal is regarded as one of the highest honours a mathematician can receive (wikipedia).

A 3D rendering of a Calabi-Yau manifold laser etched into a glass cube by Bathsheba Grossman

Maybe a 10-dimensional universe is not so farfetched after all.

Consider another belief: Gaia[2].

This is (to some) a fringe belief which constitutes Earth as a living organism. However, in the Epilogue to his book describing what a plant knows (Chamovitz, 2017), Israeli scientist Daniel Chamovitz succinctly summarises the large body of work that accords intelligence to (admittedly brain-less) plants.

Furthermore, Suzanne Simard, in her book exploring the concept of a "Mother Tree" (Simard, 2021), documents identifying the complex web of soil fungi used by forest trees to communicate with each other and identifies the existence of "Mother Trees" as entities which nurture their seedling progeny and, in a sense, conduct the symphony of life, death and regeneration occurring in our native forests.

Both are powerful advocates for believing in the concept of rational conservation which draws my own belief.

Another emerging field which engenders my intuitive belief is Quantum Biology. In their book: Life on the Edge: The Coming of Age of Quantum Biology (Al-Khalili, 2014) Jim Al-Khalili

---

2   In Greek mythology, **Gaia** is the personification of the Earth[3] and one of the Greek primordial deities. Gaia is the ancestral mother (wikipedia).

and Johnjoe McFadden present a convincing analysis (to me, at least) of the mystery of how enzyme molecules can conduct their amazingly precise and efficient catalytic action in the wet and congested intracellular environment. They describe experimental measurements in both plants and animals that suggest that in the environment of the active sites of the enzymes involved in, say, respiration, photosynthesis and the firing of neurones, quantum mechanics prevails, alone providing a mechanistic explanation for the extraordinary efficiency they exhibit in such an unlikely environment.

Again, their rational elucidation of their belief engenders belief in me.

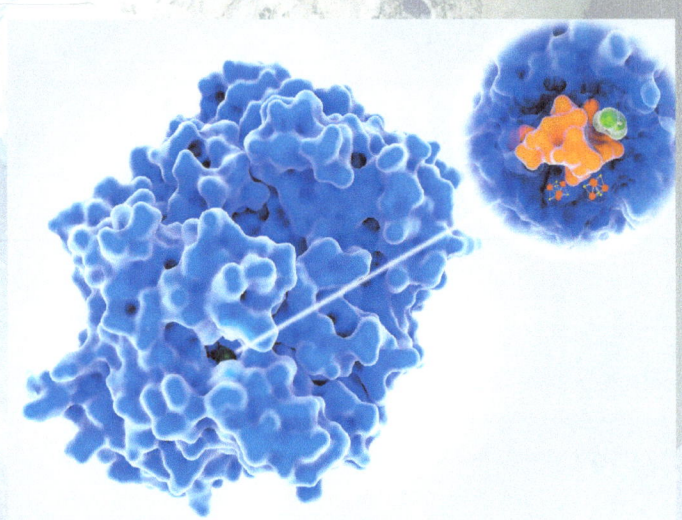

A model of an enzyme molecule highlighting the Active Site

While acquiring the information underlying the ideas presented here, to my delight, I discovered a new book, Helgoland, by Carlo Rovelli (Rovelli, 2021 (Translated Edition)).

He chose the title "Helgoland" because it is the name of the barren North Sea Island to which the young physicist Werner Heisenberg retreated during the Summer of 1925 (to escape the allergic hay fever plaguing him) whilst he thought his way through the challenge set him by his professor, Neils Bohr.

Bohr set him the task of rationalising the behaviour of electrons in appearing to orbit nuclei in precise orbits of precise energies (which Bohr had elucidated: Bohr's Rules).

Heisenberg was 23 years old at the time.

And he did it!!

Rovelli quotes Heisenberg thus:

It was around 3 o'clock in the morning when the final results of my calculations were before me. I felt profoundly shaken. I was so agitated that I could not sleep. I left the house and began walking slowly in the dark. I climbed on a rock at the tip of the island, and waited for the sun to come up...

And again:

When the first terms seemed to come right I became excited, making one mathematical error after another. As a consequence it was around 3 o'clock in the morning when the results of my calculations lay before me. It was correct in all terms.

Suddenly I no longer had any doubts about the consistency of the 'Quantum' Mechanics that my calculation described.

I am in awe of the self-belief of this 23-year-old young man that allowed him to ignore the current ideas of his elders and re-orient his thinking.

It is interesting that of the five principal explorers of this new field of Quantum Mechanics, four (Heisenberg, Pascual Jordan, Paul Dirac and Wolfgang Pauli) were in their 20's and their elder statesman, Max Born, was in his 40's.

Olivia Newton-John beside her Nobel Prize-winning atomic physicist grandfather, Max Born. Source- lavozdegalicia.es

In Gottingen they call their physics *Knabenphysik*, or 'boy's' physics!

These reminiscences anchor the first chapter of Rovelli's book. What then follows is, for me, a treasure.

PETER COCKRUM

All the learned writings I have perused have not quite accorded with my long harboured 'gut feeling' about how things work. To my absolute delight I discovered that Rovelli, in the rest of this book, elucidates what I could not.

His text, with typical scholarly excellence, provides for me the underpinning for my 'gut feeling': a *'relational'* interpretation of Quantum theory.

In Rovelli's words:

What does quantum theory tell us when there is no one measuring? What does quantum theory tell us about what happens in another galaxy?

The key to the answer, I believe, and the keystone of the ideas in this book, is the simple observation that scientists as well (as the particles being observed[3]), *and all their measuring instruments, are all part of nature.*

What quantum theory describes, then, is **the way in which one part of nature manifests itself to any other single part of nature**.

Quantum mechanics is not intrinsically human relevant, we just experience our own fragmentary glimpse of the nature of the universe. We do not actually see particles, rather we see the result of interactions of fundamental forces. We only see anything at all because our measuring instruments create these interactions. The iconic electron/double slit paradox is thus explained.

If this seems enigmatic, understand that the words I quote appear on page 67 of a 188-page book - all is elucidated if you read the book!

Key to my purpose here is to point out Rovelli's words:

"...the answer, **I believe**, and..."

To believe something is an inalienable human right! No one can take belief away from another; it is inside your head. It is an intensely personal thing **but**, neither is anyone entitled to try to force their belief on someone else. People can share their belief and attempt to draw others to it by rational argument, but they cannot require others to accept their beliefs. They cannot accord it the status of a fact.

---

3   My words

TRUTH

The definition given above makes clear that truth is far from an absolute entity: *...accepted as true,* begs the question: *by whom?* Truth, however, is of fundamental importance in human communication.

Human communication involves language and therein lies its first obstacle. Errors in translation can have consequences from comical to fatal. A comical example: in the 1980s the laboratory in which I was then working acquired a high-resolution mass spectrometer from a Japanese firm. It was installed and made functional by Japanese engineers then turned over to us to come to grips with its operation, aided by a comprehensive operating manual (in English). We quickly became bewildered by repeated instructions in the manual to "allow" something on the computer screen. One of us eventually realised we had a translation (or perhaps a transliteration) problem: traditional Japanese speakers tend to pronounce the letter "R" as an "L". When we read "allow" as "arrow" everything suddenly made sense!

Fatality is hovering when one realises that the English word "gift" means "poison" in German!

A further obstacle is exemplified by the childhood game called (in my youth) "Chinese whispers". Children stand in a line and the first child whispers a simple message in the ear of the adjacent child. This process repeats down the string of children then the final child repeats the received message out loud. The difference between the initial message and what the last child says is often hilarious.

Humorous, but the potential for miscommunication is real.

These are mechanical issues but consider the role of truth in such fields as religion, politics, and marketing. I believe the issues here involve choice and are existential for humanity.

HUMAN COMMUNICATION, RELIGION, POLITICS AND MARKETING

Over the last (roughly) ten thousand years, civilisations have emerged in many parts of our globe: Africa, Asia, South America and, most relevant for me, Europe. Around the rim of the Mediterranean Sea Mesopotamian, Greek, Roman and Egyptian civilisations emerged, developed and, like Carthage, some disappeared.

DNA analysis (courtesy of the organisation Ancestry) tells me my heritage is about half Irish, one quarter Scandinavian and the rest a mix of British, Spanish and a few other bits.

Integral to these emergent civilisations was the emergence of religions. When we humans observe something puzzling or frightening which we cannot understand or predict or control, rather than accept it *per se.*, we invent the existence of 'Gods' to whom we ascribe the power of control. Typically, the bureaucracy which grows up around the interpretation of the way these gods interact with humans (religion) becomes an end in itself, and its protagonists influence other components of a civilisation in the name of 'Religion'.

The term 'politics' is the name we give to the mechanics of the way humans organise themselves as they interact with their environment. All of us have some measure of aggression in our makeup and, over time, some of us seek and gain power over the activity of others. By and large the right to exercise that power is either usurped (Autocracy) or sought (Democracy) and gives rise to 'Governments'.

I am a child of my times and I live in a democratic society. My observation is that truth has becomes a very rubbery concept in our democracy, paid lip service but almost completely replaced by the concept of "Spin". This is very apparent if one subjects oneself to the pain of listening to what passes for debate in our federal parliament. Speakers almost never preface their remarks with *I believe* or *In my opinion* because; having banded together in groups (Parties) to maximise their chance of election, they undertake to stick to the party line and not express what they really think.

The 'facts' they present are selected from all that is known about a topic, to present one facet of that knowledge which best supports their preconceived point of view. Speakers from other parties do the same and debate largely consists not of listening and analysing, but rather trying to drown the words of the current speaker by yelling nonsense over the top of them. The farce of these proceedings is demonstrated when, after debate, a vote is taken and, regardless of the (sometimes) sensible points raised by speakers, all members vote according to their party lines anyway and the result is predictable based simply on the number of members in each party.

In this institutionalized human communication, the common human motivators of greed and the love of wielding power are supplemented by religion (every session of Parliament is opened with recital of the Lord's Prayer) and brought to bear on those we elect to office. Truth is the casualty.

The advertising industry nakedly manipulates all these behaviours; selectively presenting one facet of all the knowledge about something and engaging greed and love of power as often displayed in a lifestyle unaffordable by most.

"We have met the enemy and he is us"[4]

Fundamental to all of this is our own humanity. We can exercise choice. We choose to let ourselves be manipulated when we accept passively that these influences are being brought to bear on us and simply cannot be bothered to address them or tell ourselves that we are powerless to influence them.

Great men (Martin Luther King, Nelson Mandela…) and great scientists (Galileo Galilei, Isaac Newton, Nicolas Copernicus, Albert Einstein…) have chosen to address them and humanity has benefitted.

Religion: Some observations and reflections

As remarked earlier, religions arise as a bureaucracy associated with relating humanity to a god. They are human inventions and as such share human virtues and frailties. At the face-to-face parish level many people have been helped and supported through crises in their lives. At the Institutional level, patronage has permitted artists to produce many of the finest achievements of humanity, particularly in architecture, paintings and sculpture, and also provided the venues for its preservation (*e.g.* The Vatican).

On the other hand, I think that more blood has been spilt and lives lost over the span of human history by corruption (*e.g.*, The Borgia Popes…), and internecine wars fought to establish superiority of one religion over another (e.g., Oliver Cromwell, Henry VIII…) than any other cause.

Both sides claiming the patronage of their god (and sometimes the same god!).

---

4   First used on a poster to promote Earth Day in 1970. It is a pun on a message in which commodore Oliver Hazard Perry reported, "We have met the enemy and they are ours" to William Henry Harrison after the Battle of Lake Erie in 1812 (larrybush2013, 2014)

The 1961 British film *The Singer Not the Song*[5] chronicles the struggle between a Roman Catholic priest in a Mexican town and a Bandit Chief who is committed to removing the influence of the Church and this priest from the region.

The quality of the production gains a lot from the leading characters: two of the foremost English actors of the time; John Mills as the Priest and Dirk Bogarde as the bandit.

John Mills and Dirk Bogarde in a scene from The Singer or the Song

The bandit reluctantly admires the priest and focuses on resolving the question: is it the quality of the man himself (the Singer) or the influence of the Church (the Song) which drives the Priest in his resistance.

The film ends with a shootout between the bandits and the Mexican Army in which the bandit chief is fatally wounded. The Priest offers the Last Rites which the bandit refuses because his final assessment is that it is The Singer Not the Song which he admires.

I find myself concurring with the bandit. I believe that good people do good things because they are good at heart, not because of some supernatural influence.

A wry observation in the book Kim by Rudyard Kipling (Kipling, 1901) appeals to me. Kim, a native boy known to the locals in Lucknow as "Little Friend of All the World" gets involved with a Tibetan Lama, an Afghan (Pathan) horse trader (smuggler) and an English Colonel in the Indian Army (in charge of the Great Game -spying). In the hands of the Army, Kim's welfare is the charge of Bennett, the Army Padre (Church of England) who wants Kim knocked into shape as a good sahib by the army but is out of his depth with the boy and consults his Roman Catholic colleague.

---

5   Based on the novel by Audrey Erskine-Lindrop (Erskine-Lindrop, 1953)

Kipling writes the following: "*...between himself and the Roman Catholic chaplain of the Irish contingent lay, as Bennett believed, an unbridgeable gulf, but it was noticeable that whenever the Church of England dealt with a human problem she was very likely to call in the Church of Rome. Bennett's official abhorrence of the Scarlet Woman (Mary) and all her ways was only equalled by his private respect for Father Victor..*"

With Father Victor it is the Singer, Padre Bennett the song.

EPILOGUE

Credo (Latin: [ˈkreːdoː]; "I believe") (wikipedia)

I grew up in the Anglican communion with the Apostle's Creed used at Baptisms ("I believe..") and the Nicene Creed in services ("We believe.."). Both forms predicate belief. Reflection on the content of the Creed has led me to separate myself from the Church as an institution. I simply do not believe: .. *in one God, the Father, the Almighty, maker of heaven and earth, of all that is seen and unseen. (www.anglicancommunion.org)*

As I have previously written (Cockrum, 2020) I am open to the existence of something fundamental. Given that 94% of everything existing in the cosmos is currently identified only as "Dark Matter" or "Dark Energy", how could one deny that possibility. But I see no evidence for personification into a god – any god. This is my prerogative.

In general terms however, *in my opinion* whilst it is fine to *state* shared belief, *extending* that belief to the status of an established fact and pushing others to accept it as such, *is not*.

In the context of this work, everyone consciously prefacing any statement, especially about religion, with "*I/We believe...*" or "*In our opinion...*" removes any claim to absolute accuracy and may therefore reduce the brashness of their claims and the hostility of response by those holding different beliefs.

It may even save the World!

BIBLIOGRAPHY

Al-Khalili, J. A. (2014). *Life on the Edge: The Coming of Age of Quantum Biology*. Bantam Press.

Chamovitz, D. (2017). *What a Plant Knows: a Field Guide to the Senses (Revised Edition)*. Melbourne and London: Scribe.

Cockrum, P. A. (2020). *Getting This Far: Betrween a Planck and a Parsec*. Melbourne: Xlibris.

Erskine-Lindrop, A. (1953). *The Singer Not the Song*. Heinemann, Britain.

Kipling, R. (1901). *Kim*. MacMillan & Co. Ltd.

larrybush2013. (2014). The Morphology of a Humorous Phrase: "We have met the enemy and he is us". *https://humorinamerica.wordpress.com*.

Rovelli, C. (2021 (Translated Edition)). *Helgoland*. Allen Lane.

Simard, S. (2021). *Finding the Mother Tree: Uncovering the Wisdom and Intelligence of the Forest*. Allen Lane. Penguin Random House .

Soanes, Catherine (Ed.). (2003). *Compact Oxford English Dictionary*. Oxford: Oxford University Press.

Stephenson, N. (2008). *Anathem*. New York: Harper Perennial.

Yau, S.-T. (2010). *The Shape of Inner Space*. Philadelphia: Basic Books, Perseus Book Group.